Imagine this...

Stories Inspired by Agriculture
2018

Paisley Peterson
Dottie Davis
Natalie Gonzalez
Nathan Tanega
Caroline Thomsen
Joey Linane
Sahib Sangha

Learn About Ag
California Foundation for
Agriculture in the Classroom

Ray and Debbie
Jacobsen

© 2018 by California Foundation for Agriculture in the Classroom

Material in this publication is for classroom use and may be reproduced with credit and notification given to the California Foundation for Agriculture in the Classroom.

Publisher's Cataloging-in-Publication Data

Peterson, Paisley
  Imagine this : stories inspired by agriculture 2018 / Paisley Peterson, Dottie Davis, Natalie Gonzalez, Nathan Tanega, Caroline Thomsen, Joey Linane, Sahib Sangha
  p. cm.
  ISBN 978-0-9850855-7-5

  Summary: 2018 winning stories from the California Foundation for Agriculture in the Classroom, Imagine this... Story Writing Contest written by third through eighth grade students.

  [1. California—Social life and customs—Fiction. 2. Farm life—California—Fiction. 3. Food—Fiction. 4. Short stories. American.] I. Peterson, Paisley. II. Davis, Dottie. III. Gonzalez, Natalie. IV. Tanega, Nathan. V. Thomsen, Caroline. VI. Linane, Joey. VII. Sangha, Sahib. VIII. Title.

PZ5 .I326 2018
813.54—dd22   2018931431

California Foundation for
Agriculture in the Classroom
2300 River Plaza Drive
Sacramento, CA 95833
(916) 561-5625 • Fax (916) 561-5697
(800) 700-AITC (2482)
LearnAboutAg.org

Imagine this...

Stories Inspired by Agriculture
2018

This book is dedicated to all of the volunteers who bring agriculture into schools throughout California.

Thank you for inspiring students to #LearnAboutAg and for bridging the gap between agriculturalists and the general population.

California agriculture is special and so is the work you do.

Table of Contents

Introduction

The world of agriculture is celebrated through the creative words of talented California students in grades three to eight. The 2018 *Imagine this...* Stories Inspired by Agriculture book showcases seven unique tales about topics ranging from a day at the county fair to almonds and artichokes to even transforming into a praying mantis! The stories have been brought to life through illustrations by students in Northern California high school art programs.

Annually, thousands of students throughout California have participated in the statewide contest coordinated by California Foundation for Agriculture in the Classroom. Students select an agricultural topic, research it, and write a creative story to share with others. This contest gives them an opportunity to develop a better understanding about where their food and fiber come from and how they are produced.

The *Imagine this...* Story Writing Contest challenges students to use creative thinking and their imagination to share what they learn through writing. The stories will help children and parents understand how and where their food and fiber are grown.

We hope you enjoy the stories, and we encourage you to share your appreciation of agriculture with others.

Amanda's Dream

By Paisley Peterson
3rd Grade, Gratton Elementary School
Sheila Amaral, Teacher
Stanislaus County
Illustrated by Franklin High School

One day I was outside in my backyard and a tiny praying mantis crawled up my arm. I jumped, then screamed and shouted, "GET OFF ME!" Once I realized what it was, I zoomed in for a closer look. I was amazed that this strange, green creature could be so cool looking. As I looked at his triangular shaped head and his big googly eyes I began to wonder what it would be like if I were a praying mantis. Maybe I could do a STEM project for science class and build an invention that could turn me into a cool praying mantis.

4

"Amanda! Time for dinner," Mom called.

As I ate my dinner all that I could think about was my praying mantis buddy. "Mom, do you know anything about praying mantises?" I asked.

Mom smiled at her inquisitive daughter and replied, "Well, I know that praying mantises are **beneficial insects**. That means they are helpful because they eat harmful garden pests like **aphids**, crickets, and mosquitoes and they don't harm plants. They use camouflage as protection. I've seen one that looked like the stem of a plant."

"Wow! If I find one, can I keep if for a pet?" I asked.

As Mom tucked me into bed she smiled and whispered, "Good night. We'll talk about that tomorrow."

I quickly fell asleep and started to feel odd. I looked at my hands and I realized that I wasn't myself. I was a praying mantis! To be more specific, I was a California Praying Mantis, also known as *Stagmomantis californica*. My arms were my front legs and had sharp spines on them. I also had four other legs. My hands were like little scissor claws. I also had those big googly eyes.

Suddenly I turned my head 180 degrees and saw an aphid about sixty feet away. I couldn't resist the urge to bite the back of its neck. The aphid was paralyzed and ready to eat. Yum! That was a delicious treat.

Then a frog jumped out from behind a rock in the garden. It jumped at me, then jumped at me again! Instinct told me that Mr. Frog wanted me as a snack. I flew...wait, I fly?

I was so amazed by flying that I forgot to look where I was going and landed in a puddle. My wings were pretty wet. I quickly fluttered them to dry them off. "Oh look, a moth!" I was hungry again and was determined to get that moth. Success! I stopped to rest, looked around and I fluttered my wings again. I was finally able to fly away.

The sound of my alarm clock ringing startled me! Slowly I opened my eyes. I looked at my hands. I had ten fingers. I sat up in my bed and heard my mom calling me for breakfast. I ran to the kitchen to tell my mom about my crazy dream.

"Mom, I had the most amazing dream! I dreamed that I was a praying mantis!" I blurted as she sat down at the kitchen table.

"Really? What was your favorite part about being a praying mantis?" Mom asked smiling.

"Well, I think it would be that I had six legs and my front had spines that I used to catch and hold on to my prey. I even thought an aphid was a yummy treat," I said.

"What was the scariest thing that happened in your dream?" asked Mom.

"For sure when the frog jumped at me. I just knew that frog wanted to prey on me for a snack," I replied.

I ran to my room to get ready for school. I looked out my window and saw my praying mantis friend sitting with his legs upright, looking like he was praying. I bent down and put my face as close to the window as I could to get a good look. The praying mantis seemed to be smiling at me.

"Now I understand what it is like to be like you," I whispered to my friend, "Now when I see you in the garden I know you are a real garden helper."

The praying mantis flew off.

I watched him fly away, waved, and said, "I hope I see you again someday."

Paisley Peterson, age 9

Paisley got the idea for her story when she was sitting in her family's kitchen talking to her mom. The idea just popped into her head! She said her favorite part of the writing process was thinking about turning into a praying mantis. She hopes that by reading her story people will learn a few facts about praying mantises, just like she did while writing the story. When asked what she is most looking forward to now that she is a state winner, she said she is most excited to see her story come to life and meet the illustrators.

Franklin High School • Elk Grove Unified School District
Jessica Urban, Jasmine Lei, Isabella Hofsdal, Emma Padilla
Art Instructor: Derek Bills

Amanda's Dream was illustrated by four talented artists from Franklin High School in Elk Grove, California. The team members are Jessica Urban (senior, top left), Jasmine Lei (junior, top right), Isabella Hofsdal (junior, bottom left), and Emma Padilla (senior, bottom right). The faculty advisor, Derek Bills, assigned each artist a specific role based on strengths that he observed in class. After the team came to a consensus on the layout of the illustrations, Emma took on the challenge of creating each drawing. The inking was performed by Isabella. The watercolor was painted by Jessica. The final touch-ups with colored pencil were done by Jasmine. Each artist was excited about the project, adored the story, and enjoyed working as a team on such a memorable experience.

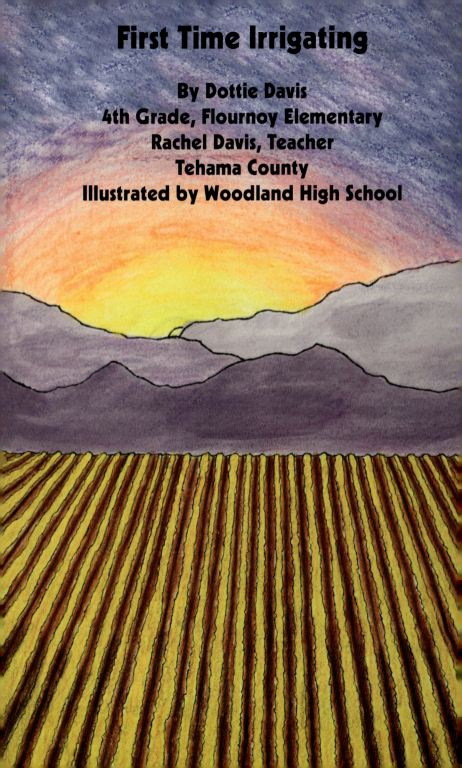

First Time Irrigating

By Dottie Davis
4th Grade, Flournoy Elementary
Rachel Davis, Teacher
Tehama County
Illustrated by Woodland High School

"Wake up! Wake up!" my dad yelled.

"But dad, it is the middle of the night! Why do I have to wake up?" I asked confused.

"Because we have to irrigate," said my dad.

"What?" I asked.

"Irrigate," my dad repeated.

"Why?" I asked.

"I'll explain later," said my dad. So I got dressed. Then, I went outside to the mule, and my dad was there waiting for me. A mule is a side-by-side vehicle, and we use it to go and check **irrigation** water.

"So, Dad, what is irrigating?" I asked.

"Irrigating is how we water our hay. Water goes down a ditch. There are little metal pieces that have a handle, and you pull the handle up. This lets the water get into the field. That is called a gate," explained my dad.

"That is cool," I said excitedly. Then we drove over to a really odd looking shape by a ditch.

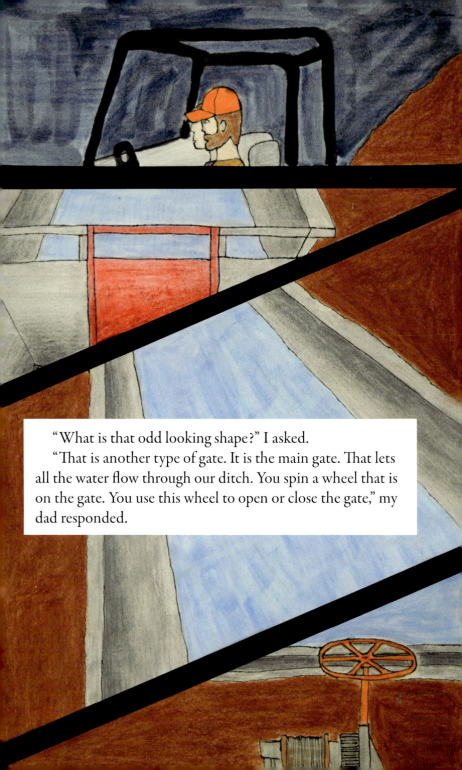

"What is that odd looking shape?" I asked.

"That is another type of gate. It is the main gate. That lets all the water flow through our ditch. You spin a wheel that is on the gate. You use this wheel to open or close the gate," my dad responded.

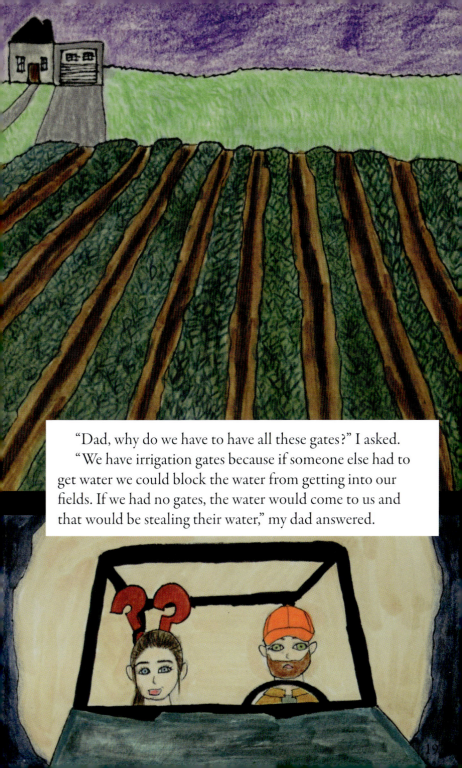

"Dad, why do we have to have all these gates?" I asked.
"We have irrigation gates because if someone else had to get water we could block the water from getting into our fields. If we had no gates, the water would come to us and that would be stealing their water," my dad answered.

"Hurry, help me spin the wheel. We only have a few minutes to get the big gate open, and all the little gates on the second field open. Wait, open all the little gates. But, only open the gates on the second field," my dad said in a rush.

So, I went to the second field, and I opened all the gates, except one. It was stuck.

I pulled as hard as I could. I pulled, and pulled, and pulled, but it would not budge. I saw the water coming down the ditch. The gate still would not move.

Finally, I pulled as hard as I could. I got it out, but I pulled so hard that I fell backwards. I got up just in time. It was a relief. I saw the water flow into the field. I ran over to my dad and told him what happened.

We have two fields. Each field has ten acres. We irrigate one field at a time, because it goes a lot faster.

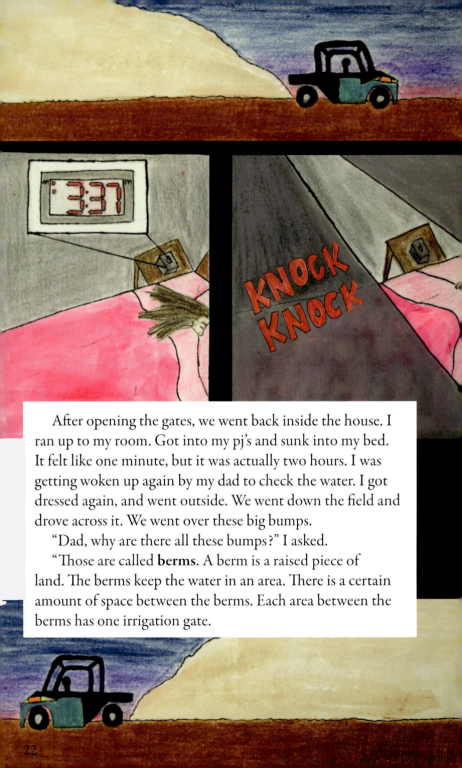

After opening the gates, we went back inside the house. I ran up to my room. Got into my pj's and sunk into my bed. It felt like one minute, but it was actually two hours. I was getting woken up again by my dad to check the water. I got dressed again, and went outside. We went down the field and drove across it. We went over these big bumps.

"Dad, why are there all these bumps?" I asked.

"Those are called **berms**. A berm is a raised piece of land. The berms keep the water in an area. There is a certain amount of space between the berms. Each area between the berms has one irrigation gate.

"Well, the water is about to the end of the field," my dad said. We drove back to the gates. "Dottie, go close the gates on the second field, and I'll open the gates on the first field," my dad told me. I went over to the gates and closed them. My dad and I got back at the same time. We went back inside. I went to sleep again.

It was nearly morning. We checked the water again. It was almost done, so we closed the gates. Then, we went to the big gate, and waited for the water to go out. When it drained, we closed it. Now, it is time to give the water to the next farmer.

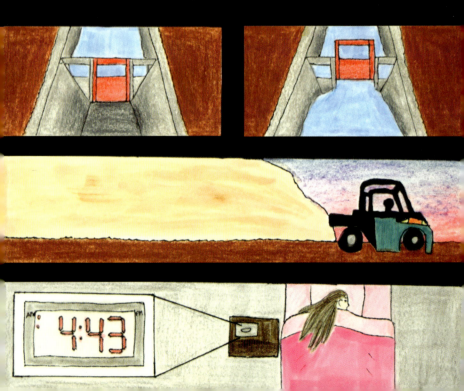

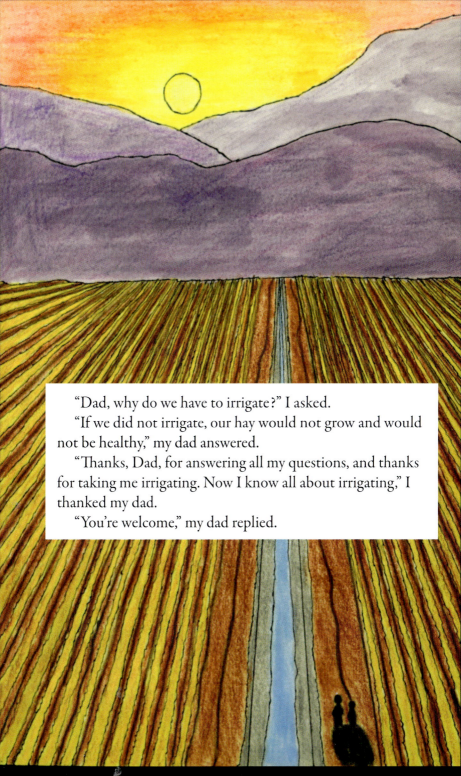

"Dad, why do we have to irrigate?" I asked.

"If we did not irrigate, our hay would not grow and would not be healthy," my dad answered.

"Thanks, Dad, for answering all my questions, and thanks for taking me irrigating. Now I know all about irrigating," I thanked my dad.

"You're welcome," my dad replied.

Dottie Davis, age 9

After contemplating two or three story ideas, Dottie thought of her topic because of her dad who is always the one to irrigate their hay. Her favorite part of the story is when the irrigation gate was stuck and she had to quickly open it, resulting in her falling backwards. In addition to learning about irrigation, she hopes that readers of her story enjoy the action! Dottie is excited to travel to Sacramento for the *Imagine this…* award ceremony and to be recognized as an award-winning author.

Woodland High School • Woodland Joint Unified School District
Megan Lacy, Lyndi Wax, and Lyneah Wax
Art Instructor: Scott Coppenger

First Time Irrigating was illustrated by juniors Megan Lacy, Lyndi Wax, and Lyneah Wax. Megan and Lyneah learned a lot of new things about irrigation, Lyndi enjoyed explaining how all the systems worked together since she already knew about irrigation thanks to several years of taking agriculture classes. They all enjoyed coming up with the idea to make the book look like a comic book because they all love superheros. (And we all know the true superheroes are farmers!) All of their watercolor skills improved. Lyneah learned what patience is, Lyndi learned she needs to let some things go, especially working with partners, and Megan learned how to paint inside the lines.

Lyndi illustrated the story in pencil first and the three worked together with watercolor paint, colored pencils, and ink to complete the illustrations. Their teacher, Mr. Coppenger, helped remind them to stay on task, but only when they needed it. They were all greatly inspired by the story in this book and are very grateful for the experience.

An Important Lesson
By Natalie Gonzalez
5th Grade, Kerman-Floyd Elementary
Falhon Ferguson, Teacher
Fresno County
Illustrated by Valley High School

This is a story about a girl's adventure, along with her friends, that teaches an important lesson about self-confi—well let's, find out, shall we?

Early in the morning Sofi the **cotton boll** woke happily knowing that it was the day of an important meeting in barn 32.

"Gooooood morning!" Sofi squealed happily.

Her mom asked, "Why are you so excited?"

"Didn't you know mommy... I'M GOING ON A FIELD TRIP! The leader of barn 32 told us," she said in a sassy voice.

"Okay, if you say so," her mom replied.

Sofi had almost blown her cover! Little did her mom know that Sofi was involved in a horrible situation. The truth was Sofi was going on a "sort-of" field trip, but not to your typical zoo or art museum. Nope. Sofi was going to their rivals over in Imperial Valley to settle her family's fate through a contest.

You see, once cotton is picked from the plant, it gets shipped out to a cotton factory, where it will no longer be soft, fluffy clouds of cotton. Instead this cotton would go through many stages and be turned into **thread**. None of the cotton around barn 32 wanted that. A little while later Sofi and her mom walked to the huge cotton barn where the leader lived. The whole cotton family was finally going to talk about a contest that was held every year.

"All right, as you all know in a week or so, the three brave cotton bolls are being sent out to represent barn 32. The cotton going are Sofi, Aiden, and Maya! Hopefully they answer the questions right so barn 32 can be saved from **processing!**" the leader bellowed.

During the two weeks, each studied like a high school student would for their SATs. Finally, it was the week of the contest. Sofi, Aiden, and Maya had just barely managed to hop on a truck that was going to Imperial Valley.

"Aren't you excited?" squealed Maya. She was super excited to show off what she had learned in the past two weeks.

Aiden and Sofi looked at each other and replied in unison, "Nope." Sofi and Aiden were not excited because they knew the fate of their family was in their hands. Finally, they arrived at Imperial Valley. The whole test was about the history of cotton.

"LADIES AND GENTLEMEN, WELCOME TO THE ANNUAL '*DO or SUFFER*' COMPETITION!" the announcer roared. The whole crowd eventually calmed down, but the same excitement still buzzed everywhere. Most of the crowd still had no idea who would be representing barn 32.

"Today we have Sofi, Maya, and Aiden representing barn 32! ARE YOU READY?" he called. Letting the suspense get higher and higher he finally started with the first question.

"Okaaay...one contestant from each barn, please come forward," he chimed. After a couple of awkward moments of silence, the sound of shuffling feet could be heard as contestants made their way to the stage.

Maya was up first for barn 32.

"Okie dokie, our first question is, when was the oldest cotton fiber and boll fragments found?"

Maya sneered, "That's the easiest question. The answer is the year 5000 BC in Mexico."

"CORRECT!" the announcer called.

Next Aiden went up! "Alrighty your question is: cotton has been worn in India and Egypt for over how many years?"

"For over 5,000 years," Aiden replied.

"Correct! Okay for our last group of constants," he hollered, "Your question is: as early as 1500 BC, cotton was worn by who?"

"The Native Americans," Sofi responded.

"Correct! I can't believe barn 32 answered first for all the questions," he prattled.

"Is it over? Did we win?" Maya chuckled nervously.

"NOPE! BONUS QUESTION! When were cotton seeds first known and recorded?"

"Urm," Sofi was drenched in sweat, she was nervous. This was the hardest question to remember. "My mom told me that her ancestors were a cotton seed that was found in 5 BC. Right?"

"I can't believe it but YES! CORRECT!" the announcer proclaimed.

"HOORAY! We got it right!" the three immediately cheered and celebrated.

They'd done it! They had won. To be honest Sofi had doubted herself so much before the contest that she didn't think that she would, or even could, win. But there they were, celebrating. They would stay happy forever. She learned when you believe in yourself, you can achieve anything.

Natalie Gonzalez, age 10

Natalie was inspired to write her story inspired by a movie, *A Second Change*, and a book, *The Accidental Hero*. Mixing the movie's theme of a gymnast on the verge of quitting and the book's theme of believing in yourself, she wrote *An Important Lesson*. Her favorite part of the writing process was writing the end of the story. "I love a good ending where the characters succeed and everyone feels proud in the end," Natalie said. She hopes that readers will learn that when you believe in yourself, anything is possible! Having self-confidence can motivate you to accomplish great things!

Valley High School • Elk Grove Unified School District
Emily Chavez and Johana Santillan-Meza
Art Instructor: Alexandra Pease

Emily Chavez and Johana Santillan-Meza were selected to illustrate *An Important Lesson* because of their amazing work on last year's story. Both students are currently enrolled in art classes at Valley High School, one in AP art and one in photography. They worked together to set up times to work and communicate about the illustrations for the story. Emily worked on most of the main character drawings while Johana worked mostly on the background and detail work. Before drawing, they researched ideas for the character's clothing, cotton textures, and colors. The artists and instructor discussed possible ideas for the story and were inspired by anime (chibi) using the cotton texture for an overall feel of the subjects in the drawings. Emily and Johana used oil paint, watercolor, and a fine point permanent marker for all the details on all the illustrations. They enjoy the process and know that this experience will lead them to more opportunities in the future.

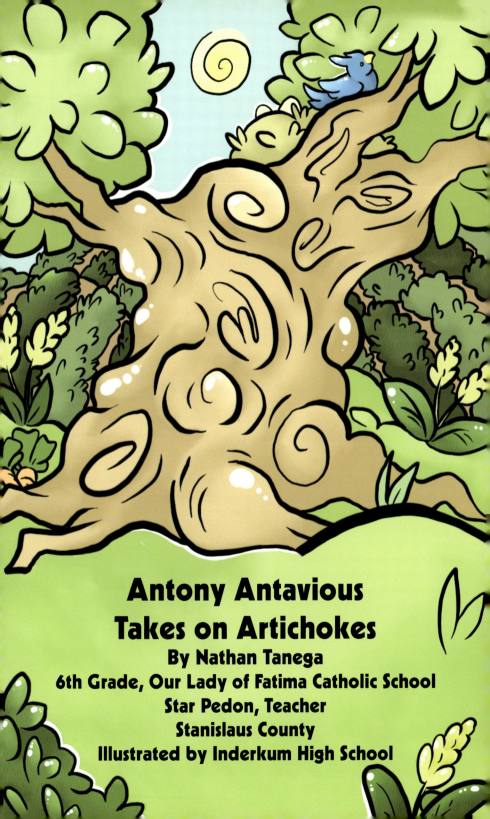

Antony Antavious
Takes on Artichokes
By Nathan Tanega
6th Grade, Our Lady of Fatima Catholic School
Star Pedon, Teacher
Stanislaus County
Illustrated by Inderkum High School

There once was a charming little family of ants called the Antavious. There was papa ant named Antony, mama ant named Antoinette, oldest son ant named Antonio, and youngest daughter ant named Antaro. They all lived happily on their little farm in ANT-ioch, California, growing their almond trees and harvesting the nuts. The family was content and enjoyed eating the almonds that they produced.

They would enjoy almond waffles for breakfast, almond butter and jelly sandwiches for lunch, and almond-crusted steak for dinner with, of course, almond cake for dessert.

They lived happily this way for many years until one summer they started to notice that the abundance of almonds was starting to decline and there was now a shortage! The **drought** had taken its toll in California and the trees they farmed no longer had enough water. This caused the almond trees to no longer produce enough almonds for their family.

Antoinette was saddened because she could no longer give Antonio and Antaro almond milk for breakfast or have their almond meals. "What are we going to do?" she cried to Antony. "The children need their milk and food!"

"Do not fear, my dear. I shall find a solution!" Antony told his wife.

So, Antony set out to find a solution to their problem. He researched all day and night and, at last, he came across a clever idea. He came across a plant that did not need water to grow! He discovered Jerusalem Artichokes, a vegetable that could grow with very little water.

Antony then talked to his cousin Antwan, who was a gardener in S-ANT-a Cruz, and confirmed with him that these vegetables grew well in most **climate zones** in California. And even better was that the artichokes could be planted in containers so they wouldn't affect the almond trees.

Antony and Antoinette learned from Antwan where to buy the tubers and started to plant them in the potting plants on their farm. In just a few months, the artichokes were growing and they were able to gather loads and loads of this new vegetable with very little watering needed.

Antoinette began cooking her family new meals with the artichokes. They enjoyed artichoke omelets for breakfast, sautéed artichokes for lunch, and artichoke soup for dinner with, of course, artichoke ice cream for dessert. The family was once, again, humming with happiness at their good fortune.

Though the artichokes were a great substitute, Antony and his family still missed their almond meals. "What can we do?" they thought. After some reflection, they decided that a prayer for rain, and a little rain "d-ANTS" never hurt anyone. So that is what they did. Every day, and every night, they would say a rain prayer, and did a little rain "d-ANTS." They did this for weeks, and was almost ready to give up.

However, one evening, after finishing their artichoke salad dinner, they were playing their board game "Smarty P-ants," when they started to hear the pitter patter of rain drops on their roof! They all scuttled outside, and to their delight, saw rain coming down!

The rain continued coming down in the coming weeks, and in no time the almond trees started to blossom once more. They were once again able to harvest their orchard of almond trees.

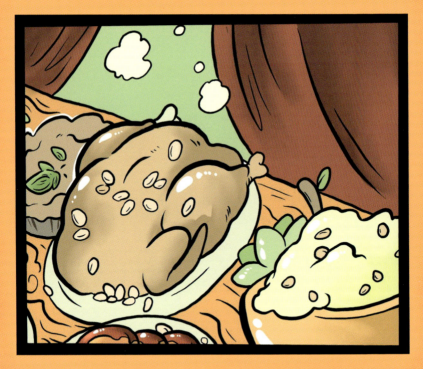

When Thanksgiving came, Antony and his family were even more grateful than usual. They looked at the feast on their table, and gave thanks for the blessings they were given. They said a prayer of gratitude before their meal of almond-crusted turkey, mashed potatoes and almonds, and yams with almond sauce with, of course, artichoke pie for dessert! All was well once again, in the Antavious ant family, and the almonds and artichokes were harvested abund-ANT-ly for many years to come!

Nathan Tanega, age 12

Nathan was inspired to write his story after his family watched the movie "Ant Man." He learned a lot about different types of ants and decided to turn them into his story's characters. After doing research about California agriculture he learned that Jerusalem artichokes are drought resistant! Nathan's favorite part of the writing process was researching names and cities that had ANT in the name. Nathan is excited to see his story come to life in the illustrations!

Inderkum High School • Natomas Unified School District
Afrah Said and Annissa Torres
Art Instructor: Rachel Rodriquez

Antony Antavious Takes on Artichokes was illustrated by students in Inderkum High School's art department. Afrah Said and Annissa Torres are both seniors. This was Afrah's second time illustrating for the *Imagine this...* project and Annissa's first. The two students worked together to create the story's unique look. Afrah focused primarily on the backgrounds and scenes while Annissa made the animated ant characters come to life. Using digital tools they merged their work together. Afrah's favorite part of the project was the illustration of the Santa Cruz skyline. Annissa said reading the story she ANT-icipated the puns using the word ant, and loved every one. Both artists are excited to see their final work come together to tell a great story!

A Fair To Remember
By Caroline Thomsen
7th Grade, Gratton Elementary School
Rexann Casteel, Teacher
Stanislaus County
Illustrated by Delta High School

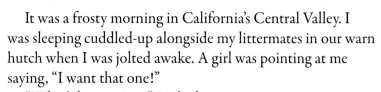

It was a frosty morning in California's Central Valley. I was sleeping cuddled-up alongside my littermates in our warn hutch when I was jolted awake. A girl was pointing at me saying, "I want that one!"

"What's happening?" I asked.

"Congratulations Clover! You're going to be Louise's show rabbit," said my mama.

"Huh?" I said, still confused.

"That girl is Louise. She needs a rabbit to compete in showmanship at the county fair. Louise had her pick of the litter and she picked you!"

My brother Cashew interrupted, "I want to be a show rabbit like Clover."

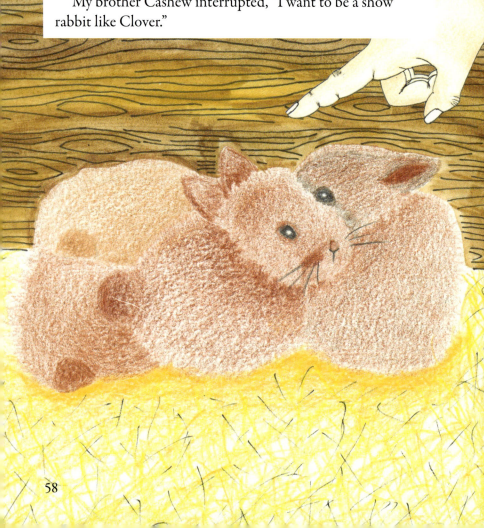

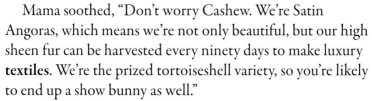

Mama soothed, "Don't worry Cashew. We're Satin Angoras, which means we're not only beautiful, but our high sheen fur can be harvested every ninety days to make luxury **textiles**. We're the prized tortoiseshell variety, so you're likely to end up a show bunny as well."

The next day Cashew and the rest of my litter were taken to the Turlock Rabbit Show. Louise visited me when she got back and said excitedly, "All your siblings went home with nice people. Some were adopted by showmanship club kids like me!"

citizenship

leadership

responsibility

life skills

"Mama, what is showmanship club?" I questioned.

"It's a club where kids learn about **agriculture**. The club helps kids develop citizenship, leadership, responsibility, and life skills."

A few weeks later, Louise took me to my first club meeting. I saw other kids with strange-looking bunnies, which I discovered were different breeds. In fact, there are forty-eight different breeds recognized by the United States! I learned a mother rabbit is called a "**dam**" and the father is the "**sire**." Females are "**does**" and males are "**bucks.**"

I was proud to learn that Angoras are the only rabbits whose fur is classified as wool, and that it's used to make scarfs, hats, and many other clothing items.

As winter turned to spring the club meetings focused on showmanship, which is where kids learned a 16-step judging routine to show and critique their rabbits. We rabbits were flipped over while they checked our bodies from nose to tail. Louise checked my eyes for **conjunctivitis** and my teeth for **malocclusion**, which is where the bottom teeth extend over the top teeth. I got wiggly when Louise checked my belly for infections called **abscesses**. It really tickled! I tried hard to cooperate because I'd grown to love and trust Louise.

When summer arrived, Louise and I practiced our showmanship routine every day. I learned to keep still during my starting pose and resisted flipping over, even when it tickled.

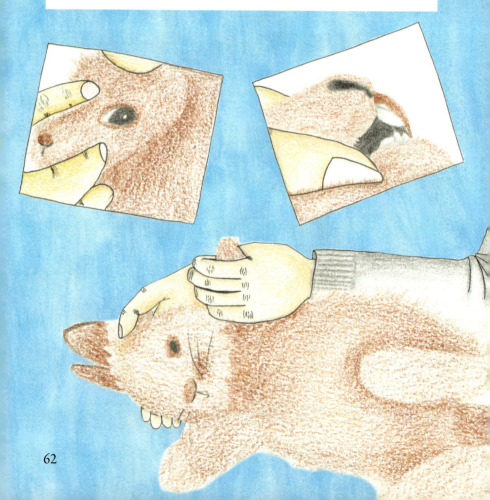

Finally, one July morning, Louise's dad drove us to the Stanislaus County Fair. Louise pointed out animal exhibits for cows, swine, poultry, sheep, and even llamas. I drooled over the prize-winning carrots in the horticulture exhibit. I overheard Louise's dad tell her, "There are seventy-eight fairs across California. Fairs help people learn about agriculture and make a positive economic and social impact on local communities."

When we arrived at the rabbit barn, Louise put me in my assigned cage and promised, "I'll be back to feed and water you every day."

In the next cage I saw a rabbit that looked familiar. The rabbit said, "Clover?"

"Cashew!" I squeaked. I was thrilled to be reunited with my brother. Cashew said he had been adopted by a kind boy named Oliver who was also in the showmanship competition.

On show day, Louise and Oliver wore matching slacks and shirts. They both sported green and white 4-H hats. Louise wore a matching green scarf and Oliver a green tie.

We walked over to the show ring and they called Oliver's name. "Good luck Cashew," I called.

"Thanks sis," he replied.

Louise and I watched Oliver and Cashew do their routine. Then Louise was called to the ring. I was so nervous! Louise set me down on the judge's table and all our training kicked in. We moved through the routine with ease and left the ring feeling triumphant.

When the competition ended, the judge awarded tenth through third place. Then the judge announced, "The contest for the top spot was very close. In second place is Louise and Clover! And in first place is Oliver and Cashew!"

"Congratulations Cashew!" I squealed.

"Thanks Clover. I bet you'll bring home the blue ribbon next year."

Louise gave me a snuggle and said how much she loved me and enjoyed our fair experience together. I beamed back at her and realized I couldn't feel more like a winner than I already did!

Caroline Thomsen, age 13

Caroline's story was inspired by her own experiences. She has two rabbits that she participates in 4-H showmanship with. Her favorite part of the story writing process was coming up with the characters, thinking what they would like, deciding their names, and what they would go through in the story. She already knew a lot about showmanship, however she did research into county fairs and how they help people learn about agriculture. She's most excited to meet the illustrators... she hopes they like rabbits, too!

Delta High School • River Delta Unified School District
Front: Oscar Romero, Nikki Martinez, Paty Perez
Back: Mike Martinez, Destiny Bettencourt, Denielle Jugal,
Tiffany Velazquez, Maddy Shively
Art Instructor: Corrie Soderlund

Eight students from Delta High School's Art III class created the illustrations for *A Fair to Remember*. The students first read through the story and broke it up by tasks. For example, one student chose to be responsible for drawing the main characters of the story; rabbits Clover and her brother Cashew. Another student focused on the various settings in the story, illustrating the fairgrounds and the rabbit barn. A third student worked on the human characters of the story. It was truly a collaborative undertaking as the remaining students read through the story and discussed which visuals were needed and whose skill set would work best with each illustration. In the spirit of collaboration, the feedback and knowledge about showing rabbits that other students in this same class period shared was very helpful to the illustrators. Through the opportunity to illustrate this wonderfully written story, the illustrators learned a lot about Satin Angora rabbits and the hard work it takes to be ready to compete in showmanship at a county fair.

The Himalayan Blackberry
By Joey Linane
8th Grade, Los Olivos Elementary School
Suzanne Squires, Teacher
Santa Barbara County
Illustrated by Sheldon High School

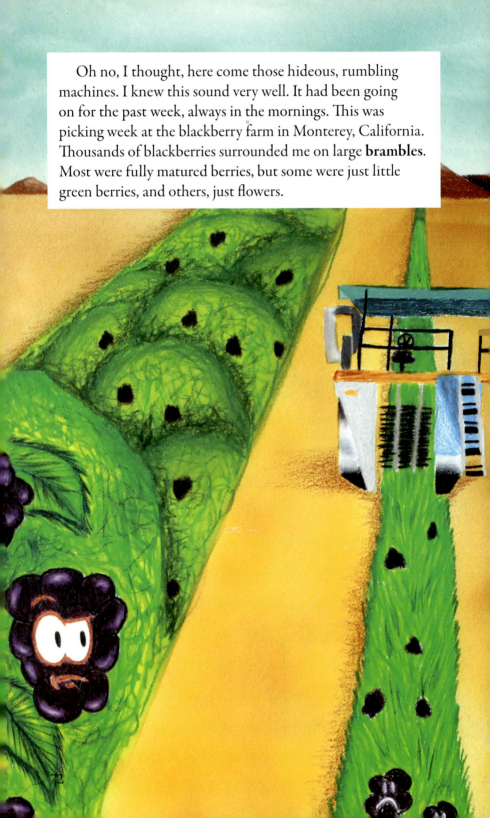

Oh no, I thought, here come those hideous, rumbling machines. I knew this sound very well. It had been going on for the past week, always in the mornings. This was picking week at the blackberry farm in Monterey, California. Thousands of blackberries surrounded me on large **brambles**. Most were fully matured berries, but some were just little green berries, and others, just flowers.

The loud rumbling of the tractor carrying the pickers spread fear among the blackberry brambles. Everyone dreaded the day in which they would be picked and packaged in plastic containers or smashed into jam, or juiced for expensive drinks. The grumbling of the tractor grew louder and louder. We couldn't do anything except pray for the best. All hope vanished for the row of brambles I was in as around fifty pickers got off the tractor and began picking all of us blackberries and throwing us into a cardboard box.

One picker quickly came up to where I was attached to the vines. I prayed he wouldn't see me but I was the first he saw. The second I got picked I expected to see a final flash of light, then nothing at all. To my surprise, I felt perfectly normal and could see everything fine. I was tossed into the center of the box while other berries were tossed next to me and alongside me. Hundreds of berries filled up the box until I could no longer see the sky. The last bit I saw looked like it was the afternoon and the picking would continue on.

It felt like hours until I felt myself get thrown onto a scale and weighed. In the process of getting weighed, hundreds of other blackberries hit me and knocked me around, leaving half my berry mush, and the other half badly bruised and red with juice. A man started sorting out the berries, either calling out "store," or "juice." I knew what would happen to me if I were to be made into juice. I hoped I would go to the stores, but quietly, in my condition, I knew I was going to a juicer to be made into juice.

"Juice," called out the sorter. I was picked up and thrown into another cardboard box along with hundreds of other damaged blackberries. Once again more damaged berries rained down on me until I could no longer see the light. A few minutes later a lid was placed over the top of the box. We were placed in a large truck and sat for what would probably be another few hours.

Four hours later we were all packed up and the truck began to drive off. Two hours later the truck came to a stop. We were lifted out of the truck and brought into an air-conditioned building. We were dumped out onto a flat tray and once again sorted, in groups of hundreds, for juicing. We were gathered up and scooped into a titanium bowl.

Right before we were tossed into the blender to sort the juice from the seeds and **pulp**, I looked up and saw a calendar reading July 25, 2018, blackberry picking season. We got dumped into the blender with the lid placed over the top and the blending began and everything went dark.

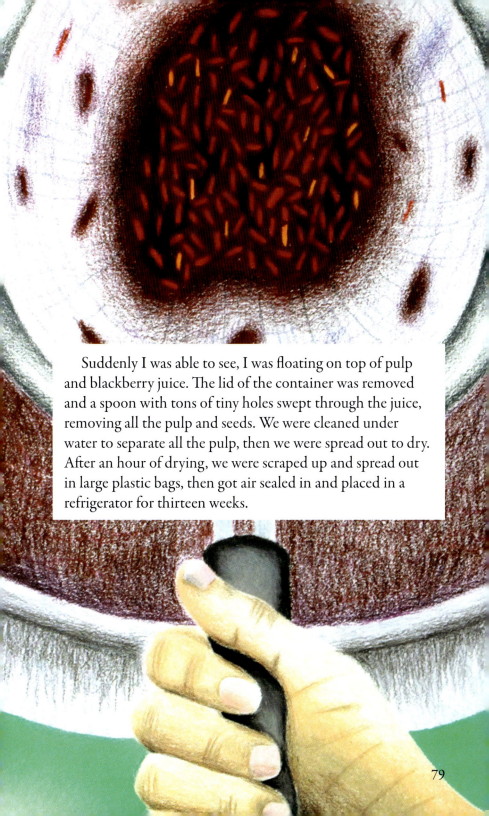

Suddenly I was able to see, I was floating on top of pulp and blackberry juice. The lid of the container was removed and a spoon with tons of tiny holes swept through the juice, removing all the pulp and seeds. We were cleaned under water to separate all the pulp, then we were spread out to dry. After an hour of drying, we were scraped up and spread out in large plastic bags, then got air sealed in and placed in a refrigerator for thirteen weeks.

When the weeks passed, we were packaged in small bags that read Himalayan Blackberry. We were driven a short distance away then removed from the package. We were on a dirt field with hundreds of rows for growing. All of us seeds were planted in the many rows, evenly spaced out. I was tossed somewhere toward the end then covered with soil and four **fertilizer** pellets.

One year later I grew into a full bramble. Another year later I began to start **fruiting**. Every three years or so the process repeated in some way and I always ended up in a new blackberry field.

Joey Linane, age 14

Joey was inspired to write about blackberries when he traveled to Oregon to view the Great American Eclipse. When he learned that berries grow in California he knew it would be a great fit for a story. His favorite part of writing his story was placing himself in the shoes of a blackberry. He hopes that readers learn that blackberries are grown and processed commercially and aren't only a plant lining the freeway. He's excited to see how the artists envisioned the settings and characters and comparing them to his vision.

Sheldon High School • Elk Grove Unified School District
Diana Herrera and Anjelica Verdin
Art Instructor: Kelsey Dillard

Before reading the story Diana and Anjelica hardly knew what blackberries were, let alone Himalayan. They learned that blackberries can be in many forms including jam, juice, and berries themselves. Diana and Anjelica split up the work about half and half: Diana illustrated five panels along with helping with the airbrush, Anjelica also illustrated five panels and helped with the coloring. To achieve the look of the berries accurately, they both had to research blackberries and many of the supplies needed to process the berries, such as the machines that picks the berries or makes juice. First, they used reference pictures to sketch from. Then they transferred their final sketches onto illustration boards and decided on what colors to use. Finally, it was time to color with colored pencils. They used an airbrush to shade in the spaces. What they liked most about this experience is being able to make a story come to life for the first time. Diana liked the fact that she could incorporate her own style into these drawings while Anjelica liked being able to draw things she was unfamiliar with. Overall, they both found the story interesting, creative, and educational.

Illustrated by Ariela Garnica
Sheldon High School

America in a Jar

Honorable Mention

By Sahib Sangha
6th Grade, Shannon Ranch Elementary
Ann Fry, Teacher
Tulare County

"*E Pluribus Unum*," my teacher stated. My head is slouched against my desk. "Today for homework you will have to write an essay or create a project to demonstrate the meaning behind these words," continues my teacher. I knew I would procrastinate, but when my teacher said, "Due tomorrow," my eyes widened.

As she talks more about the project my eyes glance off to the clock. I start my mental countdown to the end of the day. 3...2...1...Ring! Inside our class it's like a stampede as everybody rushes out.

I walk home under the Central Valley's blazing sun with my shoulders drooped and my feet feeling like lead with the thought that I already had homework on the first day of school.

As I open my front door, the phrase "*E Pluribus Unum*" keeps running in my head. I find out online that it is America's motto in Latin which means "out of many, one."

All this thinking was making me hungry. I walk to the fridge to fix myself a snack. I get myself chips and a bit of salsa that my mom had made. The salsa is incredibly colorful and just the refreshing treat I need. As I take a bite, the words "many" and "one" start circling my thoughts. A light bulb goes off in my head. I have my big idea: America in a Jar!

I grab my wallet, a few burlap shopping bags and I bike my way to the farmers market. As I look at all the stalls full of colorful, California grown fruits and veggies, I decide to make the most diverse salsa I can.

I chat with the farmers about the produce history as I shop. I pack my bag with tomatoes and find out from the farmer that they originate from Central and South America. Next, I pick up jalapeños and find that they can be traced back to Mexico. To add even more heat to the salsa, I grab a red onion and for sourness a lemon, both of which I come to find first grew in Asia. I was lucky enough to find a ripe avocado ready for slicing with roots tied back to South America. I stroll over to the fruits and pack up a handful of strawberries which were first bred in France.

I look down at my beautiful **bounty**, bulging from my burlap bags and feel blessed to be a Californian. I asked for help in finding pineapples and mangos, but a farmer tells me that I would not find any that were locally grown. With a quick trip to the grocery store for these two fruits, my shopping will be complete and I will be ready to create "Salsa the Great."

I place a huge bowl on the counter. Time to get to work! First, I chop up tomatoes and jalapeños which both contain vitamin C. I cut the tart pineapple, which strengthens your immune system. I chop up the mango, which helps to prevent cancer. Next, I add sweet strawberries for their antioxidants, lemons for their vitamin C, and red onions for their magnesium. Lastly, I add a creamy avocado for their nearly 20 vitamins and minerals. Next, I try my best to **chiffonade** the mint and cilantro picked from my backyard, both of which help with digestion. After that step, I add a pinch of Himalayan pink salt. Salt is good because it is essential for sustaining hydration levels in the body. My version of America not only looked good but tasted superb too!

The class looked dumbfounded as I stood in front of them with my jar. I hold up my jar and say "this is America the Great! America is made up of people from all over the world, bringing with them their unique language, culture, traditions, religion and new ways to understand the world. The diversity of many created one great America. This salsa represents America because it has ingredients that were all once foreign, but now most have become a part of the United States agricultural landscape. In fact, I purchased most of these ingredients from local California farmers. All of these ingredients not only bring unique flavor to the mix, but also distinct health benefits. Individually, these ingredients are tasty, but together they are spectacular, just like our country."

Proudly looking at my jar, I ask the class, "Who wants to taste America?"

Glossary

Abscesses
A swollen area within the body tissue.

Agriculture
The science or practice of producing our resources including the five F's: Food, Fiber, Flowers, Forests, and Fuel.

Aphids
Small sap-sucking insects that are destructive to plants and rapidly reproduce if not managed.

Beneficial Insects
A type of bug that provides a service (such as pollinating or pest control) e.g., praying mantis.

Berm
A raised area of ground meant to keep water in a specified area.

Brambles
The prickly branches of a blackberry plant.

Bounty
A generous offering of something (e.g., food).

Buck
A male rabbit.

Cotton Boll
The round, fluffy clumps that form cotton on a cotton plant.

Climate Zones
Division of an area designated into general climates based on temperature and rainfall.

Chiffonade
Preparation of shredded leafy vegetables, used as a food garnish.

Conjunctivitis
A common eye disease in rabbits. Often referred to as pink eye.

Dam
A female parent of a domestic animal

Doe
A female rabbit.

Drought
A prolonged period of low rainfall.

Fertilizer
A substance added to soil to increase its fertility.

Fruiting
When a tree or plant produces fruit.

Irrigation
The act of adding water to crops to help with growth.

Malocclusion
Imperfect positioning of the teeth when the jaw is closed.

Processing
Transforming a raw product into a marketable, value-added product.

Pulp
The soft, juicy, and edible flesh of a fruit or vegetable.

Sire
A male parent of a domestic animal

Textiles
A type of cloth or woven fabric.

Thread
A long, thin strand of cotton or other fiber used in sewing or weaving.

Acknowledgments

California Foundation for Agriculture in the Classroom (CFAITC) would like to acknowledge the many people who contributed to the success of the 2018 *Imagine this...* Story Writing Contest and the *Imagine this... Stories Inspired by Agriculture* book.

Many thanks to...

Imagine this... Regional Coordinators
Sandra Gist-Langiano
Mary Landau
Doni Rosasco
Jacki Zediker

High School Art Programs
Delta High School, Clarksburg
Franklin High School, Elk Grove
Inderkum High School, Sacramento
Sheldon High School, Elk Grove
Valley High School, Sacramento
Woodland High School, Woodland

CFAITC Staff
Judy Culbertson, Publisher, *Imagine this...* book
Austin Miller, Coordinator and Editor,
    Imagine this... Story Writing Contest and book

CFAITC Board of Directors
Jamie Johansson, President   Rick Phillips
Mark Dawson   Jane Roberti
Martha Deichler   Tony Toso
Debbie Jacobsen   Kenny Watkins

Ray and Debbie
Jacobsen

About California Foundation for Agriculture in the Classroom

California Foundation for Agriculture in the Classroom (CFAITC) is dedicated to fostering a greater public knowledge of the agricultural industry. For more than 30 years, CFAITC has provided educators with quality free teaching resources, along with professional development and grant opportunities. By equipping classroom teachers, after-school coordinators and other educators, CFAITC promotes student understanding of California agriculture.

CFAITC works with K–12 teachers and community leaders to help young people learn where their food comes from and how it arrives at grocery stores and restaurants. With this knowledge, students grow up with the ability to make informed choices. Through programs and resources, educators are encouraged to incorporate agriculture into lessons on math, English, science, and other core subjects to explain the important role it plays in our economy and society.

Agriculture is an important industry in California. As more rural areas become urbanized, maintaining existing farmland and feeding the growing world population becomes a greater challenge. It is important to educate students about their environment and the opportunities agriculture provides. Students can explore agricultural careers based in science, business, engineering, communication and many other disciplines.

Why Agriculture?

We are fortunate to live in one of the most diverse and vibrant agricultural production regions of the world. California agriculture plays a key role in our state's economy and in maintaining the health and strength of our society as a whole. Farmers and ranchers produce more than 400 different commodities that we rely on every day. It is essential that children—the next generation of consumers and decision makers—grow up with an awareness and understanding of agriculture's contributions to our society.

Imagine this... Story Writing Contest
Annual Deadline: November 1

The *Imagine this...* Story Writing Contest aligns with the 4C's of the Common Core State Standards—collaboration, creativity/innovation, critical thinking, and communication—while connecting students to the world around them. The contest meets the Common Core State Standards for narrative writing. Students (grades 3–8) write narratives to develop real or imagined experiences or events (Common Core State Standards, w.3–8.3) based on accurate information about California agriculture.

Each student who writes a story receives a packet of seeds and certificate of participation. Recognition is given to 48 students as regional winners and six are recognized as state winners. State-winning stories are illustrated and published as a book.

Entry Details
- Stories must be written by one student author; no group entries allowed.
- Stories should not exceed 750 words.
- Stories should not be similar in theme to winning entries from previous years.
- Submit up to five stories from each classroom.
- Each story must have an entry form attached.

Winning stories must be:
- Titled and original, creative student work
- Fact or fiction
- Appropriate for classroom use
- Grammatically correct
- Related to California agriculture in a positive way
- Typed, preferably, or neatly handwritten
- Written without reference to registered trademarks

Please refer to CFAITC's website for past state-winning stories and a detailed list of the regional and state awards:

LearnAboutAg.org/imaginethis

Entry Form

Send story and entry form to your regional coordinator (listed on back).

Student Name _____

Title of Story _____ Word Count _____

Grade _____ Number of Participating Students _____

Teacher Name_____
(print first and last name)

Teacher Signature _____

School Name _____

School Address_____

City, State, Zip _____

School Phone ( ) _____

County_____ Region Number (see map) _____

Teacher's E-mail _____

Principal_____
(print first and last name)

Superintendent_____
(print first and last name)

School District _____

How did you hear about this contest? _____

Hometown Newspaper _____

❑ Public School ❑ Private School ❑ Home School

Postmark by November 1, annually

Regional Coordinators

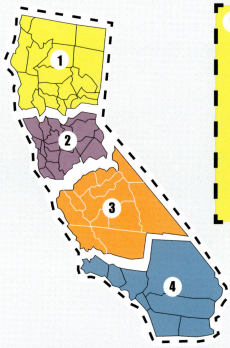

1 Jacki Zediker
7132 E. Louie Rd.
Montague, CA 96064
(530) 459-0529

Butte	Nevada
Colusa	Plumas
Del Norte	Shasta
Glenn	Sierra
Humboldt	Siskiyou
Lake	Sutter
Lassen	Tehama
Mendocino	Trinity
Modoc	Yuba

Send story and entry form, postmarked by Nov. 1, to your regional coordinator.

2 Doni Rosasco
16002 Hwy. 108, Jamestown, CA 95327
(209) 984-3539

Alameda	El Dorado	San Francisco	Solano
Alpine	Marin	San Joaquin	Sonoma
Amador	Napa	San Mateo	Stanislaus
Calaveras	Placer	Santa Clara	Tuolumne
Contra Costa	Sacramento	Santa Cruz	Yolo

3 Sandra Gist-Langiano
P.O. Box 748
Visalia, CA 93279
(559) 732-8301

Fresno	Merced
Inyo	Mono
Kern	Monterey
Kings	San Benito
Madera	San Luis Obispo
Mariposa	Tulare

4 Mary Landau
330 East Las Flores Drive
Altadena, CA 91001
(626) 794-4025

Imperial	San Bernardino
Los Angeles	San Diego
Orange	Santa Barbara
Riverside	Ventura